Edward the Alien

by

Grace Joy Warwick

Edward the Alien

Copyright © 2018 by Grace Warwick. All rights reserved.

Interior design and illustrations By Grace Warwick

Cover illustrations By Grace Warwick

All rights reserved.

No part of this publication may be reproduced or transmitted or utilized in any form or by any means, electronic, mechanical, photocopying or otherwise without the prior permission of the publisher or the author.

ISBN: 1983510734
ISBN-13: 978-1983510731

DEDICATION

I would like to dedicate this book to my Parents and Grandparents.

I would also like to thank the University of Aberystwyth for helping me to publish this book and to my parents for all their loving support whilst writing and creating this book.

CONTENTS

Acknowledgments

		Page
1	The lucky coin	1
2	Looking outside my window	5
3	Dust	8
4	As the dust rises	11
5	All that remains	15
6	Getting to the hospital	18
7	Total confusion and secrets	21
8	Making wishes	26
9	The snake filled poisoned by flowers	29
10	The fight of his life	32
11	No more wishes	36
12	False prophecies	39

ACKNOWLEDGMENTS

THE FOUR THAT INSPIRE ME...

The four that inspire me the most.
I love them all equally to the end of the earth and back and I always will.
This book is inspired by all of them, but in particular by
Edward George Warwick, the one who always saw the boring moments as fun and laughable, enjoying life to the full.
Joyce Warwick this is the only one I never knew, so I can only say what I have heard. I believe she probably had the artistic flare that flows through me. Because of her I believe I have a talent which has really matured in my art.

George Gullen, the one who confused me the most, as I was young when he left me. I only knew him a short while but it was the best time of my life.
Joy Gullen was the love that was the most complicated, erratic and wildest of them all. But also, silent, in some ways I didn't like or couldn't help, her inspiration to me, in the book was her stubbornness. It really flows throughout the voices in the book, especially in the speech of Penny. Joy's obstinate nature definitely influences Penny's to save herself, Edward and the world. I can tell that it's definitely something I inherited from her.

HOW MUCH BETTER CAN IT GET

Edward Warwick - Joyful and laughing moments.
Joyce Warwick - Never knew her, but she was the artistic influence.
George Gullen - Confused but the best time of my life.
Joy Gullen - complicated and stubborn

Edward and Joyce Warwick

George and Joy Gullen

Edward the Alien

CHAPTER 1

A round disc shape appeared in the sky. It cast a shadow over the coast of a small town in Wales called Aberystwyth. Its shadow grew larger and larger as it moved upwards, over the beach towards the people beneath it. For a while everyone stood still, amazed or shocked. On the bottom of the disc were symbols like a lucky Chinese coin shining on to the people below. The disc was a spaceship that shimmered in silver. The lucky coin symbols glistening in the light, everyone stood still looking up in awe.

"It's an alien!" someone yelled, as they grabbed their child and ran. As the terrified mother picked up her child others followed, running with her in the same direction. For one person it all happened in slow motion, he was stood there as the world passed by. His name was Arthur, he didn't know it, but he was soon to save someone's life. Little did he know he was in the right place at the right time. It was as if he knew exactly what was about to happen.

People were running in all directions, scattering and trying to escape the spaceship that had over shadowed the town. It looked bigger than everyone had thought possible. In reality it was no bigger than a bus, but it did have a huge shadow.

It began to move towards the Old College, the main attraction of Aberystwyth, hovering menacingly over rooftops. The Old College is a great part of the University and reminds people of Hogwarts from the Harry Potter books, some say it makes the town look magical. Without any hesitation, the spaceship turned this beautiful landmark into a ball of orange fire, the building started to collapse into nothing more than rubble. The spaceship struck the very top of the Old College, shattering a hole in to the tower, splintering it slowly away from the top, sending it cascading down below onto a crowd of people. Many were hurt, but none were killed. Stuck beneath the shattered tower, one man lay with a large piece of wall trapping his leg.

A stranger from the crowd pulled him out and they both ran off, clutching each other for support and holding on to each other's hand.

Two complete strangers saving each other's life.

It was a green laser that struck the tower, breaking it into a million pieces, feeding a roaring flame. People came flooding out of the Old College their screams heard above the repetitive beeping and whirling sounds of the fire alarm. They all managed to escape to see the tower slowly disintegrating.

The spaceship slowly moved over the town, sending the people below in to any and every shop or home. Afraid of what the spaceship was going to do next.

All who saw it ran, moving up the dreaded hill, the one that no one could bear to walk up on a normal. The hill was the towns main road, everyone used it, but not today! Everyone ran in the opposite direction, not knowing that what was happening on board the spaceship was even worse.

It wasn't as hi-tech inside the spaceship as you would have thought. There was no alien console or large pilot station, in fact, it looked more like the inside of a large factory. In the centre was a large room, an empty tank in the middle. it was just a glass tank devoid of any water. Around the edges of the tank were bridges with metal grated flooring and handles on the sides to stop you from falling.

Leaving this room you entered a hallway which was surrounded by smaller rooms. Some had beds, whilst others were empty, the rooms were bright white and looked like a prison cell.

Although you expected to a see pilot's cabin, there were no controls as it was piloted by someone else, someone who wasn't on board. No one knew who was controlling this spaceship, no one would know who was controlling this spaceship for another 516 years. How could anyone live that long in this story. This ship had a crew of six men and an alien, the alien that never moved. The "pilot" room looked like a large living room, a space that looked as if it wasn't being used properly. There was a small office chair in the center, facing a laptop on a desk. The same type of desk that you find in any student flat.

Next to the desk was a golden birdcage that very slightly shook. A green alien sat inside the cage shaking with fear. Scared by the six men with masks on, hiding their faces. The size of a corgi dog, it had fluffy pompoms for ears and fins like a fish. He, for want of a better word, was green all over and his hands only had three fingers, which he was using to cover his eyes. His elongated body had a similar shape to a slug, with small scales from his nose upwards. He seemed to hover in his cage, but he did not want to move. Every so often he peeped out from behind his hands to see where the men were. They never moved, they just stood there huddled, chatting in a language the little alien couldn't understand.

They were waiting for something to happen…………..

CHAPTER 2

As the spaceship moved over the town and up towards the university, it stopped and hovered over a set of flats, where many students stood staring out of the windows, not wanting to move, just paralyzed in amazement.

The spaceship stopped, hovering, between the flats, Cwrt Mawr and Rosser C. It was waiting for people to move, but no one did. Students came pouring out of the doors, they were not afraid, they were amazed, finally they were doing something interesting. They were just surprised that something this exciting was happening on campus.

The spaceship stopped outside a window, a student was woken by screaming. He stood opposite his bed, groggy and rubbing his eyes as a green light peeked through his cheap curtains, he opened them to see the outside world. Suddenly he was blinded by a green light that shone from the bottom of the spaceship towards the ground below where two girls were standing. The minute they both saw the spaceship they began to run, as it got closer the beam shot down to the ground below and they could feel themselves slowing down.

As the spaceship stopped over the top of them, it paralyzed them, freezing them midair, but they still tried to run. Frozen in mid running pose, then all of a sudden, their bodies moved. Another student stood just outside the beam, watching, trying to find a way to get help. They called the police but it was engaged, as all emergency services were handling the fire at the old college, little did they know that this was already on the news and being watched by the world.

Penny and Holly stood paralyzed inside the green beam, their skin glowing with a green hue. Their faces looked like a cartoon that was about to throw up. The man who tried to call the police, stood there trying to find a way to pull them out.

Penny felt movement in her right arm, if she could move her hand then maybe she could move her arms. She slowly reached out to the man in front of her. He put his hand under the beam, it felt like putting his hand into a pot of slime. He clenched Penny's hand not wanting to let go, but she slowly began to float upwards. He put his other hand into the beam to reach for Penny but she was being sucked up towards the spaceship. Holly was floating up and away quicker than Penny, he clenched harder he could feel the beam as it pulled them both up, getting stronger and stronger by the minute, he didn't think he could hold on to her any longer.

Slowly their hands lost grip and as their fingers brushed away from each other, Penny wondered if she knew him. This man would soon forget what he tried to do and who he tried to save, as life always gets in the way.

As she was lifted up she passed the windows of many flats where the students stood staring in amazement, realizing they could do nothing but watch. As Penny and Holly were being dragged up they both began to realise that they could move more freely. Holly tried everything she could to stop moving upwards, Penny just accepted their fate whatever it may be. Penny watched and studied the detail of the bottom of the spaceship, imagining she was the one in full control of what was happening. If she could be in control then she would tilt the spaceship and turn off the beam until she would finally feel her body hit the ground, safe and sound. But this isn't what was happening, she still held her finger and thumb just like she was holding the spaceship in her hand.

As the space ship grew she realised this was becoming more and more real. The only thing she really pondered on was how did they get here. According to science it would take aliens light years to get to earth, but what if they did not have to travel at all.

What if they were already here, when it all began, when the earth was created….. Then it all went black!

Edward the Alien

CHAPTER 3

When Penny woke up she felt metal pressing against her cheek and she could taste iron on her lips. She thought it would be painful but it wasn't. In front of her where two black leather boots, that glimmered under the light and into her eyes, looking to her right she saw Holly barely conscious. Then she turned over to have the metal now pressing against the back of her head and some of her hair fell between the holes of the floor she was lying on.

Now looking up at ceiling she saw six men slowly peering down at her. They wore white masks with the impression of faces on them. For Penny this felt like the beginning of a horror movie or an American thriller, she wondered when she would wake up from the nightmare?

She never knew who or what they were? She stared at pale white masks that were fixed to their faces. They watched her waiting for something to happen. Penny stared into their eyes which were black as the night, but one of them had two different coloured eyes, one green and the other brown, he was the only one that seemed human. She could see a guilty look in them, perhaps it was because of what he was about to do.

As Holly came to she felt scared and shaky, two of the men grabbed her and made her stand while the others grabbed Penny. One was watching, loving all the chaotic commotion. One stepped back raising a gun and pointing it at Holly, but when he fired no bullet left the barrel, instead a blue laser came out. Both Penny and Holly struggled as hard as they could in the arms of the men, kicking and scratching to get away. Penny eventually pulled herself free but it was too late, she watched as the blue laser shot across the room and it went into Holly's chest. Holly was still standing after it hit her, as if nothing had happened. But within a split second she slowly started to fade away, turning into specks of dust. Penny had managed to get free as the men gave up on restraining her. They let her fall to the ground and crawl over to the soft golden dust that she thought was her sister. She picked up a pile of the dust and watched in disbelief as it slipped threw her fingers and slowly slipped through the holes in the floor.

Edward the Alien

Pulled roughly to her feet she was held back, as the dust disappeared. She cried for her sister, but little did she know her sister was alive and well at home.

The one who stood watching walked over to Penny, he was the only one who had the nerve to disturb her. He stood and watched her be pulled up and forced to walk and said.

"Let the bargaining begin"

His voice was deep and muffled, with a guilty tone, as if he didn't want to do any of this. Penny silently swore to herself that she would never forget that voice or his eyes. Penny expected to see little green things with massive eyes on the spaceship, not human like beings with black eyes, except one, the one who seemed ashamed and had two different coloured eyes. She vowed that she would try to find this man again.

They began walking along the bridged pathway, they left the room to enter a hallway. Penny was made to stand still and her hands were tied together, while one man held a gun to the back of her head. She was told to remain quiet unless she was spoken to and was forced to walk on again……

CHAPTER 4

Penny stood as still as possible in a room that felt like it belonged in a mansion. She looked around, it was just a big dusty room, one that almost made the moment feel less scary. She was surprised how human like this place looked and felt, she had to remind herself that she was on a spaceship.

She was sat roughly onto a metal chair and tied tightly to the arms with a thick rope. She expected to see something alien, but the room had nothing remotely alien in it. Apart from one thing, Edward. She was still misty eyed from crying over Holly. The men now stood behind her mumbling, not wanting her to hear. She concentrated and that was when she heard them whisper Holly's name, that she was alive. Penny was overcome with emotion on hearing the words and let out a big heavy sigh.

"The other girl is alive and on planet Earth." Mumbled one of the men.

She now knew her sister was alive and she knew that she should try to survive. Looking in front of her she something in a golden bird cage, she glanced over her shoulder to see the men still talking and no one had a gun pointing at her anymore. The men had relaxed, more confident now Penny was tied up, no longer mumbling, but they were talking loud enough for Penny to hear Edwards name and who he was.

Before attempting to talk to the alien she could see he was afraid. He was small and green, about the size of a corgi dog, he had fluffy pompom like ears and fins like a fish, with green hands that only had three fingers which he was using to cover his eyes. His body was elongated and had a similar shape to a slug, with small scales from his nose upwards. He seemed to be floating in his cage, hovering in the same place. She tried to make him feel less scared, whispering gently to him "Shusssh". His quivering was getting louder as he was whimpering under his breath. Edward finally perked up, realising the men were no longer interested in them. Knowing they were not looking at him, he uncovered his eyes. Pulling his arms into his body they turned into fins. He had big human looking round eyes, which to Penny were almost cute. Penny could not believe what she was seeing as she looked into the cute round eyes of Edward the alien

"Edward." The alien said in a high pitched childlike voice.

Penny tried asking the alien a few questions, but every time she got the same answer. "Edward."

Each time he said his name it was pronounced more and more in perfect English, it was as if he was learning, figuring out how to speak.

Then all went silent and Edward covered his eyes with his hands once more. In that moment Penny realised the men were no longer talking and she shut her own mouth. Not in fear but, because she knew she had to. One of the men put their hand on her shoulder squeezing tightly and said "I wouldn't try talking to him if I was you. He doesn't know any of your pitiful language." With a tone that was loud and ever so old and commanding.

Penny replied sarcastically as the man continued to squeeze causing her to feel pain "No but you clearly can" she grunted wanting him to stop. He then let go and knelt in front of her, Penny felt relieved but only for a little while. He held in his hands a broken mirror he fiddled with it in front of penny and placed it on her legs.

"Look into the mirror." He commanded.

Another man's hand forced her head down to look into the mirror, but she couldn't, all she saw was her face, blurred by a misted green. Whatever they wanted her to see she couldn't see it. "What do you see?" he asked. Penny looked up, resisting the hand on her head. "Nothing" she replied.

Edward the Alien

"Exactly you see nothing, soon you will see everything", before plunging the broken mirror into her leg. Although it was small, it still hurt. It ripped her jeans and punctured the skin. Only a little blood, it will become only a small scar, one that she would never forget how she got. For a few seconds, she saw the man get up and argue with another man behind her.

"What did you do that for?" one of them said out of Penny's sight.
"You know exactly why. It has to happen", said the one who stabbed Penny. The men then swapped places, the Guilty Man now knelt before her, she felt a little faint because of the stab wound. He took out the piece of glass and stitched up the wound best he could, ignoring the fact that every time he made a stitch she felt a little more pain. She noted through her pain that he wanted to talk, she let him start , she would learn more.

"Look" he began. "I know none of this makes sense, but soon it will and when it does. You will be able to forgive us."

"How could I ever forgive you?" replied Penny.

"Because you know who I really am". That was all he said it was the last thing he said. Penny, in that moment, started to feel close to the man, she wasn't attracted to him, it felt as if she knew him. He got up, turned on the camera that sat in front of her. Once the camera clicked into life, the little red dot shining bright, all the men requested that NASA should be listening and that the whole world should listen and they did.

They stated that a team of astronauts had stolen something from them and they wanted it back. They impressed the fact that Penny and Edward would soon save the World and the Universe. This fact was true, the entire world knew, everyone but Penny knew.

For now, it was a secret, kept from her, hidden amongst the stars. This one fact will lead to a man called Arthur who had tried to save Penny's life, by doing so he became entangled in her life forever.

CHAPTER 5

Penny was left alone for a while and she managed to have a conversation with Edward. She was right he did know our language. In the small amount of time he had been on earth he had learned to communicate, but this was the first time he had used the language.
"Hey" Penny whispered.
"Hi" said the quivering alien. "What's your name again?" Penny asked, Edward remained silent for a while.
"I'm Penny, You?"
"Edward" it replied still quivering.
"Edward, I promise I am going to get us both out of here, ok!". Penny said with a sigh of relief.
When they met earlier Edward wouldn't communicate with Penny as he was afraid he would say something wrong. Scared of saying anything, just in case he got them both hurt.

The men made a video with Penny and Edward, to broadcast across the world, to every piece of technology, trying to get NASA's attention. Them or the world, that was what they had said. Penny and Edward for their stolen things to be returned or they would be destroyed. If that wasn't enough they said that if they did not get what they wanted then they would take the world.

A few hours later..........

Penny and Edward now stood before Arthur. No longer tied up and hurting only a little. They made Penny hold onto the cage Edward was in for support then chained her to the bar with handcuffs laced in gold. She heard the chain clutter to together, as they locked the cuffs.

"Please do not try anything. It's for your own good" said the man locking the cuff.
"How is any of this for my own good" said Penny with anger written across her face. They pushed her to walk towards Arthur, who looked similar to Matt Smith, he was a doctor, a physiotherapist.

Arthur threw something across the room and it landed at her feet. Penny stopped for a second, realisation hit her with moment of clarity.

"Keep walking" said the man who stabbed her, she knelt down and grabbed the object, it was a case. It looked like a large glasses case, it reminded Penny of the case her Dad kept is sports glasses in.

The men saw what she had picked up and demanded that she drop it. She just stood there for a moment, entangled in her own thoughts. Imagining handing her dad the case as he went out for his daily run, but no matter how hard she pushed the thought she could not think of her Dad. Not one bit of her could see him, she knew exactly what he looked like. He was taller than her, taller than six feet, dirty blonde hair, brown eyes. But it was like he was blurred in her mind.
So, she decided to ask why, not that she knew who exactly she was asking.
"Why can't I see him?" she said to thin air as she turned around with the alien weapon in her hand, that was once in the case.
"Penny why can't you see who?" the Guilty Man said.
She never had a chance to hear the answer over the loud bang. She never saw where the bullet came from, but looked down and put her hand just below her chest. All she saw was blood, coating her bright red top turning it dark vermillion. Her hands were covered in it, except it did not feel like blood, not to her at least. She thought blood would feel like milk on her hands. But it didn't, it felt like olive oil, then she collapsed into Arthur's arms.

He carried her, the men let them leave. They never wanted anything, they were just making sure the moment happened to become a part of history, a part of time itself.

Everyone thought when you fainted you saw the night sky, filled with a million stars. This was true, but not for Penny. For a moment she was still conscious, she heard Arthur say to a man that they teleported to earth before handing Penny over.

Then her body went into shock..........

Edward the Alien

CHAPTER 6

Penny kept slipping in and out of consciousness, she thought she was going to die, falling into a coma for 3 months. Penny thought she would see the night sky with millions of stars laced into every bit of darkness. Dreaming of a universe that terrified her, one that showed the future and the dream gave the answers to overcoming a species called *Time* and realising *Time* were the hunters.
Penny saw herself sitting on a bed holding a small crystal ball which reminded her of a film. There was a part where the Goblin King trapped the main character inside a crystal ball. That's what this moment felt like to her, like she was trapped inside a crystal ball.

Holding it in her hand twisting it back and forth, she could only see one thing inside, a coin. Remembering what the spaceship looked like, a lucky Chinese coin. For a brief moment she remembered what she saw, when she was being dragged upwards into the spaceship.

She wished for a moment that none of this had happened, wanting to go back and save people even though she knew her life was the only one that needed saving. In that moment Penny imagined herself sitting on her bed, in the block of University flats.

Penny sat there in her clothes, a black leather jacket that shined in the light incasing a blinding bright red top, with a simple grey skirt and knitted leggings. Looking at her feet for a moment noticing that there was nothing on them, she felt the soft cover press against them and imagined this was the world she wanted to be in.

Penny held the ball in her hands inspecting it for a long while, realising that she could get the coin out. She saw a slit in the glass and broke it in half quite smoothly, both halves lay on the bed. She gently took the coin out, like slipping an award out of its case. Ever so softly she encased it in her hands, that was when she knew that this was not reality and she needed to wake up. She threw her coin into the air, her life on the line, in the real world she was in surgery.

Her life was literally a coin floating in the air and who knew when it would land, then the moment changed. It went from one of fantasy, to one of shock. One second she felt dizzy like the moment you feel just before fainting. Her body was numb, she felt as if she was floating. Penny thought that it would never end, that she would feel like this is the rest her life.

Her body slowly started to regain feeling and the pain got worse and worse. Deep inside the pain was another indescribable feeling, as if her life source was seeping away. She fought the feeling and she wanted to scream and break free. Awakening 3 months later and real-world faces stared at her. Faces she had not seen for a long while. Once all the doctors and family had been, she started to slip in and out of consciousness again. When asleep she dreamt of the coin, except in this dream she had made a connection.

The coin was now an exact replica of the spaceship, Penny remembered how the spaceship in Aberystwyth looked very much like a Chinese lucky coin, causing her to wake up thinking of aliens and how she was abducted 3 months ago.

For some time Penny was worried, she did not want her dreams to come true. But no matter what she did they will always happen, Time made them happen, because this Universe is the real Universe, this Universe is Penny and Edwards.

Edward the Alien

CHAPTER 7

When Penny finally woke in hospital a mask encased her mouth, she breathed rapidly as panic kicked in. Her body still felt numb, the first face she saw was Holly's, her twin sister. The only way you could tell them apart was their hair, as Penny had dyed it bright purple. The next time she woke she was disorientated, her breathing was heavy and her heart rate rapid, but Holly came in clutching two hot chocolates, with no apparent concerns. The doctors and nurses had already explained that Penny would wake up soon, but she was still in a panic and thinking about Edward.

What had happened to him? Where was he?
Did he actually exist?

Holly placed a cup of hot chocolate on the bedside table. Penny felt tears grow in her eyes as she asked the question "Where's Edward?" Holly answered quietly and quickly showing Penny a card and saying the alien was at home waiting for her. Penny cried in relief and Holly hugged her. Penny caught up on what she had missed in 3 months, apparently not much had happened, but now was the time to get back to reality and the niggling in the back of her brain..

She did not know whether to tell the nurses and her family what she saw, she was afraid they would call her crazy. She decided to keep it to herself, perhaps it was her version of an out of body experience, apparently quite common in coma patients.

It is said that in this moment the patient's soul leaves the body temporarily and walks among the ward or hospital using their imagination to create a believable story for when they wake. The experience is unexplainable by science itself and definitely one Penny could not begin to explain.

During this time Penny's family had managed to convince the University that she could redo the entire 3rd year once she was well enough, Holly thought it was best they do this. At first Penny felt disappointed, but she soon settled on the idea of a fresh start.

Penny knew her sister inside and out and realised that Holly was hiding something from her. She needed to find out what it was, they were alone for quite a while as the doctor said they would not come in until Penny had re-adjusted to the reality. The Doctors were concerned that seeing medical staff all the time may give Penny a mental shock.

She also finally got to know Arthur who became her physiotherapist. When he came in to introduce himself Penny spurted out his name "Arthur" before he did. Penny decided to ask her sister about the secret she was keeping from her. They spoke of aliens, the Pandora's box and not just Edward or the men in masks, there were more that the human race knew about and the fact is that they had known for a while.....

Holly began to explain that of theory that we had not yet met aliens because it would take them billions of light years to get to Earth. In reality humans learned over time that they were already here, when humanity had accepted Edward needed our help, other aliens came out of hiding.

They were running from a species with many names. *"The Hunters"*, *"Time"* and *"The immortals"*. They found earth and have lived here in hiding ever since the beginning of the creation of our solar system. The big bang attracted them and now they can never leave. They are hunters and time has now bound them to this Universe. Penny for some reason could not see them, every alien to her, including Edward, looked human, almost as if they were masked.

This was all down to Pandora's box and its mythical legend, now there was a science to it, no more myth, no more legend. Penny and Edward are now bound together to find the box so that they can show the true horrors to the world that *Time* has been hiding from everyone.

They are the chosen ones, destined to be the *Protectors* of our Universe, of all the Universes. They have become immortal and as *Time* knew the unforeseeable future, that they feared, *Time* hid the secrets of Pandora' box from the *Protectors*. The secret, which Penny and Edward now knew, that they needed to find the box in order to see and reveal all aliens. The legend has it that Pandora's box unleashes a terrifying curse upon the person who opens it. *Time* is solely to blame for all the chaos and confusion all over the universe. *Time* had to be stopped, for as long as the box was shut, they were both blind to the world of aliens, they could see each other, but everyone else looked human to them.

Edward the Alien

Time, it sits on our wrist or on the wall, watching, waiting....

What if it was alive, what if it is an alien time. History is in the wrong order because of them, they know all of history, past, present and future, they change time to suit them. Only keeping alive the people and aliens that serve their purpose. If someone or something does not fit the picture of their perfect universe, then they change history so that person or moment does not exist. They have forgotten the consequence of their actions, too many changes have occurred, keeping every universe in a cage and they are getting smaller. Time is running out and the damage that *Time* has done is the ultimate blackhole, slowly sucking existence itself into nowhere.

Once Holly explained her secret, Penny felt she should share her out of body experience, no more secrets, no more lies. Holly listened quietly as she recalled her dreams and visions. A few days later Penny was fit to leave the hospital and return home and start getting back to University with Edward. She needed to start again, a whole new beginning.....

CHAPTER 8

A few months later at the start of a new semester, a whole new beginning, just what Penny and Edward needed. Edward had made the decision to join her, even with him by her side Penny was worried as time was racing forward getting closer to what Penny had seen.

Penny often found it difficult to cope physically and emotionally, it made her worry about Edward. One night Penny was trying to shower, it was difficult since she was partially paralysed when she was shot and there was some damage to her spinal cord. She had physiotherapy and was working hard to build her strength, but it was difficult to stand for any longer than 5 minutes. Somedays she had to use crutches, depending on how her legs felt. Penny managed to shower with a stool, but she had not fixed it to the floor properly. One day, Penny went to sit down and the stool slipped out of the shower she landed on the floor in a heap and her leg was numb, all she could do was sit in the shower and yell for Edward.

She managed to reach up and pull the towel off the hook on the door just before Edward entered the bathroom to help. Once she had been lifted up to her feet and got changed, she decided that it was time for dinner. The brief crazy moment in the bathroom had made Penny realise that Edward needed to know everything and besides these dreams she feared were just interrupting her life, perhaps Edward could help with that.

They sat in the student flat, eating the pizzas that had just come out of the oven, the exact flat Penny lived in when it all began. Outside, on the streets, last year when she was being sucked up into the belly of a spaceship. It all seemed a bit weird, she found herself discussing the problems of the future with the only alien she could see.

Now they felt very close, almost like brother and sister or best friends. Pizza now eaten, crusts on the plate, Edward knew Penny's secret, no more secrets left except one. Edward decided to tell Penny about his secret and the reason why the men in the masks took him.

Edward the Alien

What he told her sounded ridiculous, she found herself wanting to laugh. When she heard Edward's childlike voice explain that he could grant wishes which were infinite and could quite literally be for anything. No rules to abide by, just as many wishes as you like and he had no choice but to grant them, no matter what the consequences maybe.

He made Penny swear never to tell anyone, if it became known what he could do he would have to start running and hide from the world. Edward has already left his species, moving from the seas and on to land to somewhere secluded. He didn't want this power but he knew because of it he had to hide, it was his responsibility to keep it safe. Then he had a burning desire to ask Penny one last question and to Edward this one was the most important.

"Do you want a wish?" he said nervously but instinctively, his way of working out how to trust someone. If a wish was asked for, then perhaps that trust was broken or was never there.

Penny sat there quietly for a while, then broke into laughter. Edward was confused at first, thinking he had missed a joke, he couldn't understand why Penny found it so funny.

Then she sat there looking at Edward and said "Seriously".

For a brief moment Penny was not thinking, she was remembering one of her dreams about this exact moment. She made a wish that she could just see where people go after they die, she had always said that she was an agnostic and placed most of her belief in science, but often wished there was some kind of person or being to look over everyone, everyone needs something to believe in.

So, she decided.

"No." she said with a smile on her face. Edward let out a big sigh of relief and knew he could trust her. Penny decided she would never want a wish, she liked her life just the way it was.

CHAPTER 9

Edward now understood why Penny was afraid all the time, but he was more concerned about who was after him. He was worrying more and more as each day passed. Penny kept saying that the men in masks were coming back someday and they would take them both. He feared this as he knew exactly why the men wanted him, whatever the wish they wanted. He knew it would be bad but learned to hide and use distractions. Edward had learned to live alongside humans and other aliens on this planet, speaking English helped him communicate. Humans now had to adjust to the fact that some of their family were aliens and that they had evolved alongside us. History was changing, but was it for the better or worse.

That future was coming faster than they thought.

Penny and Edward come home from shopping, settling back into Uni life, Penny needed to get some art equipment for her art project. She was learning to do illustrations for her favourite book, The Time Machine by HG Wells. She liked this book because these days she related to it, reminding her that knowing the future often meant you would want to change what happens.

Whilst out shopping they topped up the food, Edward went to the kitchen to put it away. In the meantime Penny organised the art equipment. Setting it up to get ready to work, she found a package lying on her bed, with a note from her sister, she was staying for the weekend, for work. The note said " I found the package outside the front door" with no idea who it was from.

Putting her work to one side she immediately opened, ripping it apart, peeling back the bubble wrap, to find a key ring and a key. For a while she seemed baffled, she went into the kitchen to find Edward, he was quietly talking to himself busy loading shelves. Penny stood there in silence watching him. The door shut slowly behind her, Edward hardly noticing her presence until he heard the door click shut.

"I thought you were sorting out your work?" he said.

"I was…" replied Penny, but her thoughts had already drifted. Standing there inspecting the key and keychain, setting them both down on the kitchen counter to get a closer look. She realised the key looked familiar. It was an exact replica of the key she received for her 21st birthday. She ran to get her flat keys and placed it on to the counter next to the other key.

Why would someone replicate a fake key?

The replica key felt heavy, this key, what did it open? Pandora's box? At least now she had figured out one part of the puzzle, now all they needed was the box. Now she had to tell Edward, because having this key meant the men in masks would return soon. He was still mumbling some of the English he had learnt, oblivious to Penny wandering around.

"Edward will you listen please" Penny yelled to get his attention. He looked up with an expectant gaze waiting for Penny to tell him. She started to explain, blurting it out that they had to stop the masked men from making the wish or else there wouldn't be a home or a universe to live in. Figuring out who the masked men were was the most difficult, Penny could only remember the one who seemed remotely human, the Guilty Man, with two different coloured eyes. She always remembered him. The keyring seemed to have meaning for a moment, thinking what it could symbolise. She began unlocking the secrets of the key in her mind. The snake was the guilty human and the flowers were time, but they are poisonous. It was once harmless and true to nature, nurturing everyone and everything, it found the flowers and thinking how delicious they looked decided to eat them and indulge himself. Once the flowers were consumed the snake became their slave and the poison of the flowers became its venom. The snake once true to nature now serves the flowers, poisoning everything it once loved. Deep inside this reptile, it knows what it is doing is wrong, but he cannot stop itself. The flowers underestimated the other animals who grew to love the snake and no matter what happened they would release the snake from the flowers power. The snake may have been poisoned by the flowers but one day the poison will be removed, whatever it takes.

Edward the Alien

CHAPTER 10

Penny was determined to find out who the Guilty Man was, she had time to figure this out. She tried the internet watching a colossal amount of conspiracy theorists on social media. They knew nothing, every time she heard their words, she heard the word theory and that's exactly what it was a theory, not truth.

She had to dig deep into her memories, especially when she was alone, mentally recreating the moments when the Lucky Coin spaceship arrived.

She lay silent on her bed, trying to imagine the faces of the masked men. Anything that could tell her who they were, all she could recall were their black eyes. The only exception was the Guilty Man, she still believed he was being forced to do their bidding.

She focused her mind on the Guilty Man.

"Black clothes" she said to herself attempting to recollect the moment she nearly lost her life. Moments before, she was looking at him, her mind wandered to her family. She thought if she could remember his face, then maybe she would be able to change that time and save her planet from the wish that *Time* wanted.

Penny had found out that the wish and many of the new alien species she had discovered wanted to tell her what they knew about *Time*. *Time* wanted infinite power and control overall the Universes, changing every time stream to suit their own needs. Entire planets to be erased from history, eradicating all their inhabitants. Rewriting history would turn any world into a never ending paradox of hell.

With this despairing thought she concentrated even more on her memories, knowing that if she could change that moment and keep Edward safe, it may save the Universe. She reached deeper into her mind for the words she had heard the Guilty Man say.

He said " You know me" Penny remembered feeling that they were related, which was not possible. She discarded that thought and information as irrelevant. She remembered him stitching up her leg and seeing his dark brown hair, protruding from the edge of his mask, possibly a beard. His hair was greasy like he hadn't showered in months. She grasped at a face from her imagination, thinking this was the right face, she drew it. His eyes held the most detail in the drawing. Golden specs of dust that *Time* had left behind when Holly had been sent back to Earth, the dust seemed to feel real. It could have been the aliens true form, shapeshifting and moving between bodies, controlling whatever and whoever they wanted. This would give them the power to travel anywhere even between universes.

Then Penny remembered the key to Pandora's box. If she opened it, then they would be one step closer to saving our Universe, perhaps even every Universe…..

But who knew what the consequences would be?
Interrupting her thoughts Edward burst in shouting "Have you seen the news."
"What?!" Penny exclaimed confused.
Edward turned on her laptop and put on BBC news. All he had to say was "They are here" and Penny knew exactly what he meant.

The spaceship was back, causing havoc, trying to get Penny and Edwards attention. In her dreams all went wrong, but now she knew what to do, she had to see his face. She went outside, Edward tried to stop her, she didn't just want to save her family or the town she wanted to stop *Time*, she had finally found the courage to save the Universe from them.

Edward managed to stop her at the door thinking he had lost her trust. She stood there looking into his eyes and said "I have a plan" and winked at him, he knew then that he could still trust her, Penny knew what she was doing. They both went outside ready to stop the chaos. Whatever the plan was, it would be for the better. As they stepped outside the street door they didn't have to yell to get the aliens attention as the green beam found them straight away and hovered above them. Instead of floating up slowly like before, it happened so fast that when they were both on the ship they were dizzy.

Penny looked to her right still a little dazed and saw Edward tied up, suspended over a large tank filled with water. This was no ordinary tank as it was full to the brim with gigantic fish, it was the last of Edwards species. He did not want to go home, did not want to join them, he did not fit in, he was not free.

Penny was being held by two masked men with dark black eyes, she knew they were *Time*. In front of her stood the Guilty Man, this time he was not in control of his words or his body. He was standing still but shaking, fighting whatever was controlling his body.
He started to speak. "I wish…" he began, the moment he did Penny began yelling. Louder and louder…….
"No, No, No!" repeating till he finished, so that Edward could not hear the wish and could not grant it.

Once the Guilty Man finished, the men in the masks threw Penny to the ground and spoke. "I wish…" they began, but Penny got up repeatedly shouting the word "NO!" until they stopped. When they had finally finished, they heard Penny say "I know, I know who you are." They both stared at her in shock and amazement.
"Who are we?" they said at the same time in one voice.
"*Time*." Penny answered.
"How do you know this?" their eyes began turning from black to golden.
"I do, because you really underestimated us!" Penny said with new found confidence.

Meanwhile, Edward was swinging on the rope in an effort to break free, he heard the rope creak as if was ready to snap. Just as Penny finished her sentence he smashed into the two masked men.

Penny ran towards the Guilty Man, tore off his mask, he seemed old, with a dark brown beard and brown greasy hair. As the masked men were getting up the Guilty Man said "It's not supposed to happen this way". "Exactly." Penny said turning around to face *Time*.
"I know who you are as well". She said pointing at the Guilty Man behind her.

Edward the Alien

CHAPTER 11

Penny and Edward stood before the enemy of the Universes and together made a wish.

"I wish we were both home".
"I wish my species were home safe and sound and all the species on this spaceship are home where they want to be and where they feel safe" Edward carried on, but Penny interrupted him saying "Don't finish the wish, this is all we need." Edward stopped as he trusted her. They watched as they saw *Time* stand there for a moment shocked, frozen, then the bodies faded and turned into specks of golden dust. Penny and Edward turned to face the Guilty Man, as now she wanted the truth, she knew that was the only way beat *Time*.

"Hey" Penny said to the man who was reeling with shock, " You are my brother?". She smiled and said "Sorry but we need to be fast. I need to know if they are watching".
"No" he said forcing the words out as the wish was trying to force his body to go back to *Time*, but he was fighting it.
"Ok," Penny said " where is Pandora's Box?"

"In the black rucksack, over there" he replied. Penny and Edward turned to see the rucksack peeking out from behind a door, they found inside it a battered looking box, where the Guilty Man had hidden it. Penny threw the rucksack onto her back, clutching the straps, never wanting to let go, they turned and faced the Guilty Man once more.

"How do we open it, when do we open it?" Penny asked.
"There are two boxes, you must open the first in three days time, not including today. The second one doesn't have a key." As he spoke Edward saw his species disappear, turning into gold dust, returning home to the seas of Planet Earth, where they were safe.

"The box will whisper to you, to open it earlier, don't or else *Time* wins." The Guilty Man carried on oblivious to what was happening around him. "Keep the box out of sight, in a place that only you and Edward know of. Keep the key on your person at all times. Also…" he could feeling the wish growing stronger, trying to pull him away.

He had to concentrate on staying to finish his conversation.

"Arthur is not your friend he cannot be trusted, *Time* has taken control of his timestream. *Time* will make an older version of him to show you the consequences of opening the box. You have to keep it away from him and ignore them, everything he does or says is a lie. Survival is the key, it may be slow but you can win and tell nothing but the truth to anyone and everyone. Lastly, don't tell them..." he stopped short, trying his hardest to resist the wish. "Don't tell them about our mother, they want to control you."

Penny and Edward assumed the man meant *Time* wanted to control Edward and Penny's time stream changing their history. "Why?" Penny asked. "I don't know, just don't tell them or else they will change your entire history, they will make sure you never met, I think they are afraid of both of you". He stopped talking and gave into the wish. He was slowly leaving but Penny and Edward would never forget his face.
"Wait" Penny said, pulling a necklace out from underneath her shirt, she had the lucky coin in her hand. She placed it around his neck, it was on a piece of black rope.
"Hide it under your shirt where *Time* won't see it." As she said those words tears filled her eyes she did not want him to leave.
"They probably won't let me keep this." He replied.
"You managed to hide that rucksack didn't you" she said knowing he would do it. Then his skin began to turn to golden dust and blew away into a cold wind that came from nowhere.

"One more thing" Penny said hoping for an answer, hoping the wish would slow down. "What is your name?" She asked.

Penny and Edward stared into the eyes of the Guilty Man, as his face faded into the wind and they heard him say "Frankie"

They looked at each other, staring at their hands as they too turned to gold dust and faded into the cold wind. On their way home to planet earth, the wish had worked.

Edward the Alien

CHAPTER 12

Day Zero
Penny and Edward stood where they were taken from their flat, their friends were behind the door, still inside frantically trying to get it open. They finally did, Holly was there and hugged Penny the minute she saw her. Penny could feel the rucksack still on her back, she was relieved that she still had it, even more relieved than she was to be home.

"You seriously think I would let them win" she said to Holly, who was squeezing her hard.

Penny and Edward had been gone for two hours, they thought they would return to a place of devastation, but they didn't, *Time* did nothing but blow up one building, the old college. They had run out of moves and were hoping that the wish would give them everything they needed. The old college was being rebuilt, construction had started to bring back a landmark that everyone remembered.

Penny felt sad for the people and aliens lost on the construction site, many had lost family members today, but they had to carry on and follow her brother's instructions, then they would be one step closer to winning.

They filmed a message to the world "I have it" Penny said "I have the Pandora's box and I'm going to open it. When I do, *Time* will be losing. People and aliens of earth I need you to carry on as if nothing happened today. A man is coming and he cannot be trusted, if we tell you to do something then it must be done. We are keeping the box as far away from him as possible. It must be opened on the 8th of March, if we don't *Time* wins." she felt the pressure of the Universe weighing on her shoulders.

"We must win."

Then they uploaded the video to YouTube, Facebook and any other social media you could think of. It didn't take long before the video went viral and ended up on the news. They all just sat there waiting in the kitchen, waiting to see what the occupants of Planet Earth would do. Everyone replied Yes or OK, the response was immense. They followed Penny and Edwards instructions because all they wanted was their time and universe back to normal.

Day One
Penny and Edward kept the box inside the black rucksack and stuffed it inside a pillow, then placed it in a cupboard above their wardrobe, out of sight.

The whispering had begun, every time Penny or Edward heard it they thought about a similar box under the bed. They picked it up, opened it, every time with a sigh of relief when it was empty. Arthur turned up, but it was the wrong one, every time they saw him they asked him how old he was, 24 was the only answer they got.

Day Two
Penny awoke with a fright during a nightmare, the boxes were whispering to her, tricking her into opening Pandora's box too early. She rolled over and tried to dream of something else, imagining opening the box with flowers inside. Drifting back off, ignoring the whispers in her ears, for tomorrow would be the day that they open the box for real, bringing them one step closer to winning.

This time Edward was also awake "Hey" she said to him "Nightmare?" "Yeah" he replied getting underneath the covers with her. They both lay there discussing the nightmare, Edward's was that *Time* had got him and was making him wish. Suddenly, Penny wondered why she had never thought of this before, she leant over towards him and whispered "Edward, I do have a wish" and before Edward could respond she said "I wish that you or any of your species would never be able to grant a wish ever again". Edward was truly free and would never fear the nightmare again, his species were no longer in danger. He smiled as he drifted off into a dreamless sleep, no voices, no nightmares. Penny fell asleep shortly after, knowing that Edward was safe.

Day Three

This is when *Time* thought the end for them would begin, they did not know there were two boxes to be opened. Penny and Edward woke at 8am, they were expecting something to happen straight away, but nothing did. They got up, ready to face anything that would try to stop them. Penny moved a chair to get the box out of the cupboard, whilst Edward watched the door. As she was reaching into the cupboard to get at the box, they heard a sharp knock at the door both looked at each other questioningly.
"Who is it?" Penny said calmly.
"Arthur" was the terse reply,
"How old are you?"
"Who needs to know". As Arthur wouldn't answer the question she assumed it was *Time* and ignored it. Penny pulled the rucksack out of the pillow and took the box out. She gave it to Edward, jumping off the chair and closed the rucksack swinging it onto her back.

Maybe Edward could give Penny some time, by distracting Arthur. She opened the window and said to Edward "Go to the old Castle ruins and wait for me there". Edward ignored every desire he had to stay with Penny and help her instead, he hovered out of the window and left.

Penny opened the door to Arthur "Where is it?" he said violently pushing the door and knocking her to the floor. He stood in the middle of the room looking for the black bag. Penny tried to stand as quickly as she could "It's not in here" she said, she whispered to herself "Come on chase me", she pulled the door open and began to run, it was a good thing that they had kept the box in a safe place. Edward and Penny had been determined to keep the box in Aberystwyth, as they knew all the short cuts and alleyways.

She ran down the steps in front of the Arts Centre slipping on a bit of ice, it had snowed earlier that week. She grabbed the nearest person "Hey can you do me a favor and go on ahead and make sure those cars stop.". Arthur was slow, he was following Penny's footsteps, she was determined to make it difficult for him. When saw him stood at the top of the stairs Penny began running again. She held the hand of the stranger and crossed the road, cars sped up and pulled in front of

Arthur, making sure he could not follow. Penny let go of the stranger's hand, "thank you" she yelled as she kept on running, the stranger smiled back, he knew she was the girl who would save the Universe. Penny kept on running.

She reached the Old Castle ruins and stood in the wind reunited Edward. She turned around to come face to face with Arthur, starting to open the box with Edward. Someone stopped her in her tracks, it was herself?

"This is what will happen if you open that box" it said. The other Penny looked like her but was much older, slightly taller and looked like a soldier. Arthur seemed to trust it even when it pulled out a gun and pointed it at him. "If you open that box, then he is lost and this is what you become." Edward handed Penny the box and whispered "Remember what your brother said, they are lying".

Penny couldn't help but let tears well up in her eyes, she did not want to lose a friend, but with that last thought she pulled the key out from under her top. Edward grabbed onto the key with her and they put it into the lock.

Penny shouted "Showing us the future makes us want to change it, all this, what your showing me, will never happen."

They turned the key and the box opened.

To be continued....

Edward the Alien

ABOUT THE AUTHOR

Her parents always told her to follow her dreams. Whilst writing this book she was studying for a BA Honors in Art and Education. Drawing had always been Grace's favorite thing to do, she never leaves the house without a notebook and pencil in her bag.

FORTHCOMING...

This book is part 1 of a trilogy.

Book 2 - Edward the Alien - Unleashing *Times* Secret
Penny and Edward have opened Pandora's box, but have they won the battle. Are Frankie and Arthur finally free from *Time*s power. What is inside and how will it help them defeat *Time*.

Book 3 - Edward the Alien - History was wrong.
Penny is 543 years old, she has a family, do they finally defeat *Time*. They must find a way to open the second box, but how? and who will help them?

RECOMMENDATIONS

If you are interested in reading another book. I have also illustrated a story called Alien Ants by Lee Warwick (My Dad). It is available to purchase on Amazon as an e-book and or paper back. Proceeds of this book are donated to the Children in Need charity.

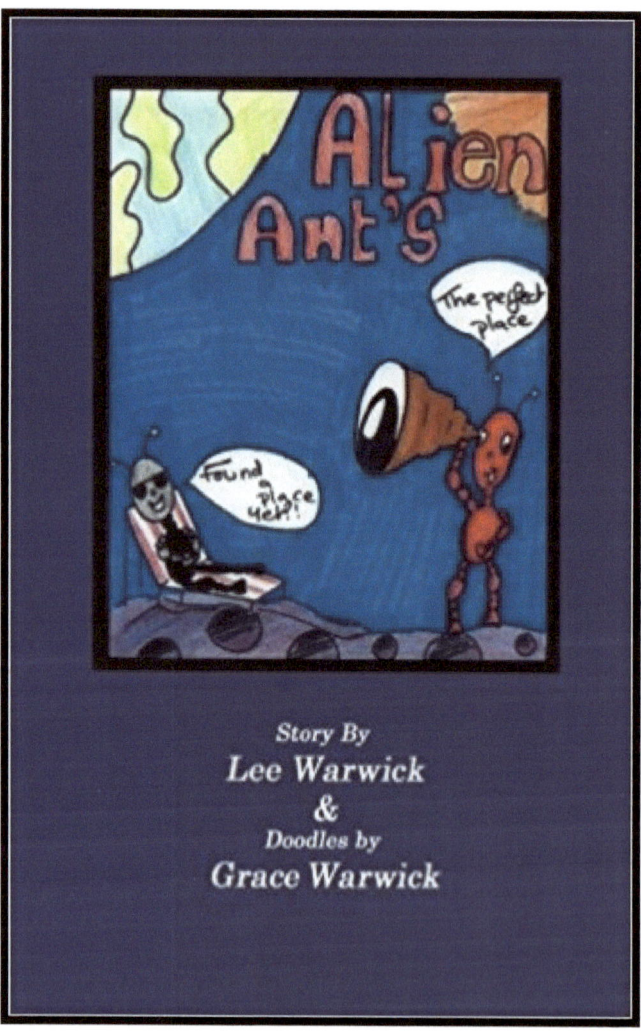

www.ingramcontent.com/pod-product-compliance
Lightning Source LLC
Chambersburg PA
CBHW040241220526
45473CB00001B/320